X-TREME FACTS: SCIENCE

PLANTS

by Catherine C. Finan

BEARPORT PUBLISHING

Minneapolis, Minnesota

Credits:
Cover, © Dmitry Brant/Creative Commons 4.0, © Sashkin/Shutterstock, and © Neil Bromhall/Shutterstock; 6 top, yakonstant/ Shutterstock; 4 top, Chansom Pantip/Shutterstock; 4, stefanophotographer/Shutterstock; 5 top, Dmitry Naumov/Shutterstock; 5 bottom, Maksym Deliyergiyev/Shutterstock; 5 bottom left, Chris Christophersen/Shutterstock; 5 bottom middle, LightField Studios/Shutterstock; 5 bottom right, Nick Pecker/Shutterstock; 6 bottom, Linas T/Shutterstock; 7 top, Jeff Holcombe/Shutterstock; 7 top right, Dieter Ebert/Creative Commons; 7 middle, Henri Koskinen/Shutterstock; 7 bottom, Egon Zitter/Shutterstock; 7 bottom right, Asier Romero/Shutterstock; 8 top, Alex Polo/Shutterstock; 8 middle, MIROFOSS/Shutterstock; 8 bottom, Dudarev Mikhail/Shutterstock; 9 top, digitalienspb/Shutterstock; 9 top right, Gelpi/Shutterstock; 9 middle, Banprik/Shutterstock; 9 bottom left, Eder/Shutterstock; 9 bottom middle, Sergej Razvodovskij/Shutterstock; 9 bottom right, colin13362/Shutterstock; 10 top, Christian Vinces/Shutterstock; 10 bottom, cpaulfell/Shutterstock; 10 bottom left, sasimoto/Shutterstock; 11 top, Alex Farias/Shutterstock; 11 top right, Daderot/Creative Commons; 11 bottom, M.Moira/Shutterstock; 11 bottom left, Happy Together /Shutterstock;11 bottom right, Bildagentur Zoonar GmbH/Shutterstock; 12 top left, Aleksey Shyshkin/Shutterstock;12 top right, Daderot/Public Domain;12 bottom, nnattalli/Shutterstock; 13 top, Mark Byer/Shutterstock; 13 top right, Ondrej Prosicky/Shutterstock; 13 middle, John L. Absher/Shutterstock; 13 bottom, Aki_spring/Shutterstock; 13 bottom left, USFWS/Creative Commons;13 bottom right, Lillian Tveit/Shutterstock; 14, evenfh/Shutterstock; 14 bottom left, Roman Samborskyi/Shutterstock; 14 bottom right, Prostock-studio/Shutterstock; 15 top, Joshua Avramson; 15 bottom, hermitis/Shutterstock; 15 bottom left, Gelpi/Shutterstock; 15 bottom right, Cristina Romero Palma/Shutterstock; 16 top, moj0j0/Shutterstock; 16 top right, pfly/Creative Commons; 16 bottom, valzan/Shutterstock; 17 top, Joshua Avramson; 17 top left, l i g h t p o e t/Shutterstock; 17 bottom, Nancy Tripp Photography/Shutterstock; 17 bottom left, Pedro Lira Rencoret/Public Domain; 18 top, NayaDadara/Shutterstock; 18 bottom, Galen Parks Smith/Creative Commons; 18 bottom middle, Kamenetskiy Konstantin/Shutterstock; 19 top, Kevin Friery/Shutterstock; 19 top right, Anna Kucherova/Shutterstock; 19 middle, Ruud Morijn Photographer/Shutterstock; 19 bottom, Wmpearl/Creative Commons; 19 bottom left, Marek Tr/Shutterstock; 19 bottom right, aSuruwataRi/Shutterstock; 20 left, Nickolay Stanev/Shutterstock; 20 right, Allard One/Shutterstock; 21 top right, Kararkadeep/Creative Commons; 21 top, Paul Hamilton/Creative Commons; 21 top right Golden Pixels LLC/Shutterstock; 21 middle, Forest Service photo by John Zapell/Public Domain; 21 bottom, Paper Street Design/Shutterstock; 21 bottom left, altanaka/Shutterstock; 21 bottom right, svitlana-ua/Shutterstock; 22 top, Guenter Albers/Shutterstock; 22 top right, Artiste2d3d/Shutterstock; 22 bottom, Benjitheijneb/Creative Commons; 23 top, Yiyo Zamorano/Creative Commons; 23 top right, Virrage Images/Shutterstock; 23 bottom left, Gorodenkoff/Shutterstock; 23 right, Lucky-photographer/Shutterstock; 23 bottom, Klokeid/Public Domain; 24 top, Zbigniew Guzowski/Shutterstock; 24 left, 24 right, Poggensee/Shutterstock; Grayson Schmidt/Shutterstock; 24 bottom, Ramjchandran/Shutterstock; 25 top, TB studio/Shutterstock; 25 top left, AlexandrMusuc/Shutterstock; 25 top right, 28 bottom, irin-k/Shutterstock; 25 bottom, SingjaiStocker/Shutterstock; 25 middle, michaeljung/Shutterstock; 25 bottom left, Alan Poulson Photography/Shutterstock; 25 bottom right, spaxiax/Shutterstock; 26 top, MSPhotographic/Shutterstock; 26 bottom left, LightField Studios/Shutterstock; 26 bottom right, Ruslan Huzau/Shutterstock; 26 bottom, photoPOU/Shutterstock; 27 top, Tomas Florian/Shutterstock; 27 top right, ch_ch/Shutterstock; 27 middle left, Kalcutta/Shutterstock; 27 bottom right, andscha/Shutterstock; 27 bottom left, Roman Samborskyi/Shutterstock; 27 bottom middle left, WAYHOME studio/Shutterstock; 27 bottom middle right, studioloco/Shutterstock; 28 top left, Diana Taliun/Shutterstock; 28 top right, Wolfgang Filser/Shutterstock; 28-29, Austen Photography

President: Jen Jenson
Director of Product Development: Spencer Brinker
Editor: Allison Juda

Developed and produced for Bearport Publishing by BlueAppleWorks Inc.
Managing Editor for BlueAppleWorks: Melissa McClellan
Art Director: T.J. Choleva
Photo Research: Jane Reid

Library of Congress Cataloging-in-Publication Data

Names: Finan, Catherine C., 1972- author.
Title: Plants / Catherine C. Finan.
Description: Minneapolis, Minnesota : Bearport Publishing Company, [2021] | Series: X-treme facts: science | Includes bibliographical references and index.
Identifiers: LCCN 2020035420 (print) | LCCN 2020035421 (ebook) | ISBN 9781647476762 (library binding) | ISBN 9781647476830 (paperback) | ISBN 9781647476908 (ebook)
Subjects: LCSH: Plants--Juvenile literature.
Classification: LCC QK49 .F45 2021 (print) | LCC QK49 (ebook) | DDC 580--dc23
LC record available at https://lccn.loc.gov/2020035420
LC ebook record available at https://lccn.loc.gov/2020035421

For more information, write to Bearport Publishing, 5357 Penn Avenue South, Minneapolis, MN 55419.
Printed in the United States of America.

Contents

The Power of Plants

Plants live all over the world—on land, in fresh water, and beneath the sea. They can survive in scorching deserts, frozen tundra, and dim ocean depths. Some live for thousands of years, and others grow more than a foot a day! Plants make their own food, and some actually trap and eat animals! Let's journey into the world of plants and explore some of the most extreme living things on Earth!

There are over 360,000 known plant **species**.

***Welwitschia*, a desert plant, relies on fog for its water. Yet it can live for a thousand years!**

A type of moss can survive in volcanic areas of New Zealand where the soil temperature is 162 degrees Fahrenheit (72 degrees Celsius).

Plants that grow near this heat have short roots that spread out sideways to escape the scorching volcanic soil.

Plants can survive on the northernmost point of land in Greenland and live in the harsh conditions of southern Antarctica.

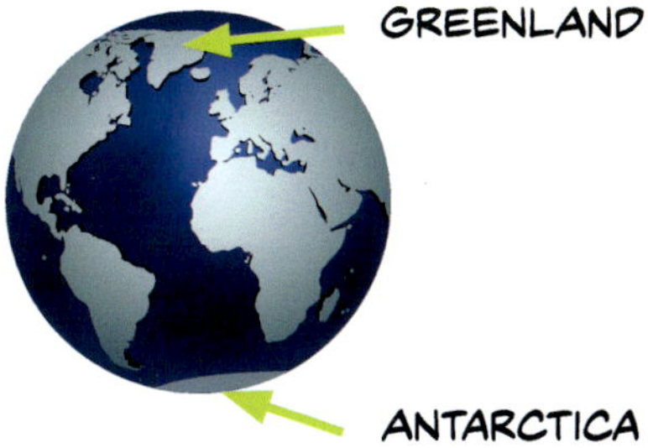

Crusty, leaflike lichen plants are so hardy that scientists think **they could survive on Mars!**

A Feast of Flesh

When we think of plants, we often think of the potted plants in our homes. We water them, and they just sit there. But some plants are a lot more . . . active . . . at feeding time. Flesh-eating plants, such as the Venus flytrap, don't wait for a meal to arrive—they actually hunt for their food. Let's look at the plants with a taste for flesh!

When prey touches the tiny hairs on the flytrap's leaves, the trap snaps shut, and the insect can't escape.

A Venus flytrap can tell time! If something touches it twice within 20 seconds, the plant knows the intruder is something alive—and yummy!

Bladderworts have an appetite for worms, water fleas, and insect **larvae**.

Bladderworts have special bladders that open like trap doors. Once prey gets inside, the doors slam shut.

The sundew produces a sticky **nectar** on its leaves. Insects come by for a treat, get stuck, and are **digested**.

The pitcher plant feasts on insects, frogs, lizards, birds, **and even rats!**

Curious critters fall into the pitcher plant and can't climb back up its slippery walls.

Danger: Poison!

Some plants may not go so far as to trap insects and eat their flesh, but they can still be quite deadly. Over time, plants have developed defenses to keep hungry plant-eating insects away. Believe it or not, the most common defense among plants is poison. But these poisons can also be harmful to animals and humans. Some plants can even kill you!

The doll's eye plant looks scary . . . and it is! The whole plant is poisonous, especially its berries, which look like freaky dolls' eyes!

Abraham Lincoln's mother died from white snakeroot poisoning when she drank the milk from a cow that had grazed on it!

The giant hogweed's leaves and sap contain chemicals that cause painful blisters.

Giant hogweed is no small matter. It can grow up to 18 ft (5.5 m) tall!

The seeds of the rosary pea are used in jewelry making, but they can kill with a simple prick to the skin!

Belladonna was used by Italian women as a **cosmetic**, but ancient Roman soldiers used it to poison their arrows!

Ouch!

Traps and poisons aren't the only things you need to look out for in the wild world of plants. Some plants cut with knife-sharp thorns and needles! For example, the puya plant in South America **impales** prey on long, spiked leaves. The animals will **decompose** on the plant, putting **nutrients** into the soil that helps the plant grow. That's gotta hurt!

Some puya plants grow to 33 feet (10 m) tall—that's about the height of two giraffes stacked on top of one another!

On the African savanna, the whistling thorn acacia tree grows sharp 3-inch (7.5 cm) thorns, which cut animals that come too close.

The poisonous needles of **the Australian gympie-gympie stinging tree can cause intense pain that lasts for years!**

The red-legged pademelon, a small wallaby, is one of the very few animals that can eat the leaves of the gympie-gympie.

Even blades of grass can cut you with their sharp edges, which are **serrated** like a bread knife.

The leaves and stems of stinging nettles are covered in needlelike hairs that inject an acid into the skin. *Ouch!*

Pollination Power

Thorns, needles, and poisons help some plants keep creatures away. Other plants aren't quite so defensive. In fact, many plants are downright inviting, producing sweet nectar to attract **pollinators**. There are a whole bunch of cool creatures that get plants growing! The chocolate midge is a tiny fly the size of a cupcake sprinkle. But without this important pollinator, there would be no chocolate in the world! Power on, pollinators!

Even pesky mosquitoes are pollinators.

The less-than-welcome wasps, beetles, and slugs in your garden are actually helpful pollinators.

There are about 2,000 species of birds that pollinate plants worldwide.

Many hummingbirds grow bills and tongues that are the perfect length and shape to fit inside their favorite flowers.

The hummingbird moth has a superlong tongue for sipping nectar. It collects and carries **pollen** as it feasts on flowers.

Over 500 plant species, including bananas and mangoes, are pollinated by bats.

Ew . . . That Smells

A sweet smell attracts pollinators to some flowers. But other flowers smell . . . less than sweet. In fact, they can even stink like rotting flesh. That's the case with the corpse flower, which is named for its disgusting odor. The world has its fair share of stinky plants. Let's sniff around and learn about some more!

The corpse flower has the tallest bloom in the world! One was 10.2 ft (3.1 m) tall!

The corpse flower may smell dead. But it **can heat up to 98°F (36.7°C).**

The carrion flower smells like rotting flesh.

Growing in Thin Air

All plants—whether thorny, needled, poisonous, or stinky—have to be rooted in soil, right? Well . . . no, actually! Plants called epiphytes (EHP-uh-fyts), or air plants, can grow in all kinds of places besides the ground. They attach themselves to a host plant for support and start growing out instead of down. Epiphytes get the water and nutrients they need from rain, the host plant, and even the air itself. That's wild!

Most mosses and some ferns are epiphytes, and they might be growing right outside your door!

What's so special about moss? This excellent epiphyte was one of Earth's first land plants **about 470 million years ago.**

Many kinds of orchids are epiphytes. Their spongy roots take in water from the air!

The rare monkey orchid has a flower that looks just like a smiling monkey's face!

Spanish moss is neither Spanish nor a moss. It is a flowering plant **native** to the Americas.

Invaders

Air plants gently attach themselves to hosts and get their nutrients and water from the air without causing any harm. Other plants, however, invade new territory and take over, pushing out native wildlife. In 1876, a Japanese plant was introduced to the southeastern United States. Farmers began planting kudzu as a grazing crop for animals. Big mistake! The plant began taking over at lightning speed. Once plant invaders creep into an area, it's very hard to stop them!

English ivy slowly kills trees by blocking their light. How rude!

The kudzu vine grows up to 1 ft (30 cm) per day.

Golden bamboo grows 30 to 40 ft (9.1 to 12.2 m) tall. When planted where it doesn't belong, it crowds out native plants.

Just one purple loosestrife can produce up to 2.7 million seeds in a year!

The Amazonian water hyacinth covers lakes and ponds, blocks the sunlight, and changes oxygen levels in the water, killing other plants and fish.

Living Large

Some invasive plants can grow as much as a foot a day and spread over a huge area. But they are not the only plants that can go big. Imagine a tree so huge you could drive a car through a hole in its trunk! Or a leaf you could lie on. What other enormous plants are out there to discover?

General Sherman, a giant sequoia, is the largest known living single-trunk tree on Earth.

The Chandelier tree has a hole cut through its base that is 6 ft (1.8 m) wide and almost 7 ft (2 m) high.

A California coastal redwood called the Hyperion tree is just over 380 ft (115.8 m) tall!

The banyan tree has the broadest **crown**, stretching out for acres.

Older banyan trees grow roots from branches that eventually turn into thick, sturdy support trunks.

A colony of 40,000 quaking aspen trees in Utah all share a single root system. That means they are actually a single **organism**!

The leaves of giant water lilies are so big that a person could lie across one with several feet to spare!

That's Ancient

A plant towering high in the sky or spreading out far and wide can be very impressive. Even more amazing may be the amount of time a plant can live in the spot where it was planted. In eastern California, a bristlecone pine named Methuselah has lived for 5,000 years. It's older than the Egyptian pyramids! What other plants on Earth have stood the test of time?

The oldest-known living plant is a huge patch of seagrass in the Mediterranean Sea that may be 200,000 years old!

Bristlecone pines are the oldest living trees on Earth. They can survive in harsh conditions and at high altitudes.

A 2,000-year-old seed from the extinct Judean date palm was planted in modern times, and the tree was revived!

An alerce tree named *Gran Abuelo*, which means great-grandfather, is 3,622 years old!

King Klone, a creosote bush in California, **is 11,700 years old!** It makes new plants by **cloning** itself.

Giant sequoias can live to be over 3,000 years old.

Talking Plants?

If ancient bristlecone, alerce, or sequoia trees could talk, imagine the stories they could tell! Talking plants—it sounds impossible, right? Yet, though they may not use words, some plants communicate with each other and with helpful insects. A tobacco plant that is being munched on by a hornworm caterpillar calls in some help from the caterpillar's enemy to get rid of the nibbling nuisance. Problem solved! Let's listen in on what some plants are saying.

When chewed, the tobacco plant releases a chemical that mixes with the caterpillar's spit and attracts the attention of the caterpillar's hungry enemies!

Threads of fungus link the roots of different plants so they can talk and share nutrients. It's a plant internet!

Corn plants send out chemicals to call on wasps, which will protect corn from other bug invaders.

Plants That Heal

While plants may call on animals for protection, humans call on plants to save us, too. In fact, our lives on Earth depend upon them! Plants put oxygen into the air, which helps all animals live. Plants give us food to eat and materials that provide shelter. In addition, many plants have the amazing ability to heal us when we are sick! It's impossible to imagine a world without plants—and who would want to?

More than 10 percent of the World Health Organization's list of **essential** drugs are made from flowering plants!

Turmeric is a powerhouse herb that is thought to keep joints healthy.

At the first sign of a cold, many people turn to echinacea (eh-kih-NAY-shuh), an herb made from the purple coneflower.

The 5,000-year-old flaxseed may keep your heart healthy.

Aspirin, a pain and fever reliever, **contains something made from the bark of the willow tree!**

Some powerful modern **painkillers are made from the poppy plant.**

Colorful Leaf Print

Craft Project

There are hundreds of thousands of plants on Earth, and each has its own unique look. The leaves of different kinds of trees and shrubs, for example, come in many shapes and sizes. Let the natural world outside your window inspire you to make your own colorful planty print!

What You Will Need

- Leaves
- Acrylic or tempera paints in bright colors
- Paintbrushes
- Construction paper

Evergreen leaves look the same year-round. Deciduous leaves change color in the fall.

Step One

Take a nature walk and collect some different leaves. Try to find leaves of different sizes and shapes. Remember to only pick up fallen leaves! If you can't get leaves from outside, find some fake plants at your local craft store.

Step Two

Using acrylic paint and a brush, paint the front of a leaf. Cover the leaf evenly with a thin layer of paint. Be careful! They are delicate.

Step Three

Lay the leaf on the construction paper and gently rub the back of the leaf without making it move around. Carefully pull the leaf from the paper. Repeat with other leaves. Let your print dry completely.

Step Four

Check out your print! What kinds of shapes and patterns do you see? Try this again with different colors and plants to make different prints inspired by nature!

Glossary

bladders sacs in some plants that can contain gas or air

cloning making an exact copy of something

cosmetic a product, such as lipstick, used to improve the appearance of part of the body

crown the branches and leaves at the highest part of a tree

decompose to decay, or rot away

digested broken down into materials that can be used by the body

essential necessary for survival

impales pierces with something sharp and pointed

larvae the stage of life for insects after they have hatched from eggs and before they change into adults; the singular of *larvae* is *larva*

native originally from a place

nectar a sweet liquid made by plants to attract insects

nutrients things taken in by plants and animals to help them live and grow

organism an individual living thing, such as a plant or animal

parasite a plant, animal, or fungus that lives on or in another living thing, called a host; a parasite gets its food and energy from the host

pollen the fine, yellow powder made by a flowering plant

pollinators creatures that collect pollen from one plant and give it to another plant of the same kind so the plant will grow

savanna a flat plain covered with grass and few trees

serrated tooth-edged or notched

species a group of living things that are organized by similar characteristics

wallaby a small- or medium-sized animal similar to a kangaroo

Read More

Lawrence, Ellen. *Extreme Trees: And How They Got That Way (Plant-ology).* New York: Bearport Publishing, 2015.

London, Martha. *Pollinators: Animals Helping Plants Thrive (Team Earth).* Minneapolis: Abdo Publishing, 2020.

Spilsbury, Louise, and Richard Spilsbury. *Killer Plants (Pilot. Engineered by Nature).* Minneapolis: Bellwether Media, 2017.

Sprott, Gary. *Plants Are Hunters! And Other Strange Facts (That's Wild!).* Vero Beach, FL: Rourke Educational Media, 2020.

Learn More Online

1. Go to **www.factsurfer.com**
2. Enter "**X-treme Plants**" into the search box.
3. Click on the cover of this book to see a list of websites.

Index

About the Author

Catherine C. Finan is a writer and plant lover living in northeastern Pennsylvania. In the summertime, she loves to watch all kinds of colorful creatures have a pollination party in her backyard garden—the hummingbirds are her favorite!